Reading

EXPLORING

The Earth

SUSAN GLASS

PERFECTION LEARNING®

Editorial Director: Susan C. Thies

Editor: Mary L. Bush

Design Director: Randy Messer

Book Design: Emily J. Greazel

Cover Design: Michael A. Aspengren

A special thanks to the following for his scientific review of the book: Kristin Mandsager, Instructor of Physics and Astronomy, North Iowa Area Community College

Image Credits:
©NASA/Roger Ressmeyer/CORBIS: p. 9 (bottom); ©Galen Rowell/CORBIS: p. 19 (top); ©Jan Butchofsky-Houser/CORBIS: p. 19 (bottom)

Comstock Royalty-Free: p. 21; Corel Professional Photos: front cover (main), back cover; MapResources: pp. 8, 18 (bottom), 19 (center); Perfection Learning Corporation: pp. 4, 5, 10 (bottom), 13, 14, 15 (bottom), 18 (top); Photos.com: front cover (bottom three), pp. 6, 7, 9 (top), 10 (top), 11, 12, 15 (top), 16, 17, 20, 22, 23, 24

Perfection Learning® Corporation
1000 North Second Avenue, P.O. Box 500
Logan, Iowa 51546-0500.
Phone: 1-800-831-4190
Fax: 1-800-543-2745
perfectionlearning.com

2 3 4 5 6 7 PP 12 11 10 09 08 07

Paperback ISBN-10: 0-7891-6602-x
ISBN-13: 978-0-7891-6602-9
Reinforced Library Binding ISBN-10: 0-7569-4645-x
ISBN-13: 978-0-7569-4645-6

Contents

A Marble in the Galaxy

If you could rocket into outer space and look down at the Earth, what would you see? From space, the planet looks like a big blue marble. The marble is spinning as it glides around a star (the Sun). The Earth and the Sun belong to the Milky Way Galaxy. They share the **galaxy** with seven other planets and billions of other stars.

Third Rock from the Sun

Planets are large bodies of gas or rock that travel around the Sun. The planet Earth is 93 million miles from the Sun. That makes it the third-closest planet to the Sun.

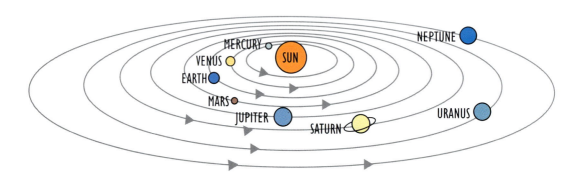

Earth is the fifth-largest planet. If you walked around the entire planet (at the equator), your trip would be almost 25,000 miles long.

The Earth has one Moon, which is about a fourth of the size of the Earth. This Moon is the only body that **orbits** the Earth.

What Goes Up Must Come Down

If you toss a ball into the air, what happens? No matter how high you throw it, eventually it will fall to the ground. Why? Because of **gravity**.

Gravity is the force that pulls two bodies in space toward each other. The forces of gravity between the planets, the Sun, and the Moon keep everything in the galaxy in its place.

Gravity pulls all objects on Earth toward the center of the planet. Gravity is the reason that you hit the ground when you fall off your bike. It is also the reason you can walk around on the Earth's surface instead of just floating away.

Technology Link

Sir Isaac Newton is the man who discovered and defined gravity in the 1600s. The scientist also had several other important ideas about motion. Using these laws of motion, modern inventors were able to create rocket ships.

Rockets use fuel to create a force that launches and moves the ship. The first rockets were unreliable. Many exploded or didn't stay on course. Advanced equipment and **propellants** make today's rockets faster, safer, and more accurate. From Newton's early ideas came an invention that enables humans to explore space in ways that Sir Isaac may never have imagined.

On the Surface of Things

Why does the Earth look blue instead of green or red or purple? Because almost three-fourths of the surface of the planet is covered by water. Oceans, rivers, lakes, and streams flow across the land, making it the "blue planet."

The other fourth of the Earth is covered by interesting land features. Tall mountains, low valleys, flat farmland, and sandy deserts spread out across the planet.

Breathe In

The Earth's **atmosphere** is a mixture of gases. Nitrogen makes up 78 percent of the air. Oxygen makes up 21 percent. The remaining 1 percent is a blend of a few other gases.

Planet Life

Earth is the only planet known to support life. The temperature, water, air, and sunlight on Earth make it a suitable home for plants, animals, and humans.

A Ball on a Stick

Pancake or Basketball?

It might sound strange to question the shape of the Earth. Today it is well known that the Earth is round like a basketball. But years ago, that wasn't true. In fact, thousands of years ago, most people thought the Earth was as flat as a pancake.

When you think about it, it's actually quite easy to think of the Earth as flat. When you look out the window of your house, the Earth looks flat. When you're driving down the road, the Earth seems flat. Even when you look out across the huge ocean, the planet appears flat. Long ago, sailors feared that if they sailed too far out in the ocean, they might fall off the edge of the world!

A few people disagreed with the early belief that the Earth was flat. The ancient Greeks assumed the Earth was round. One famous Greek thinker named Aristotle was well known for believing that the Earth was a **sphere**. One of his reasons for this belief was that at times the Earth cast a circular **shadow** on the Moon.

Some sailors recognized that when they saw a ship or island come into view, they saw only the top of it first. As the object got closer, more and more of it appeared. Eventually the whole thing was visible. If the Earth was flat, the men would have seen the whole ship or island from top to bottom when they first spotted it. It would have looked tiny, but all of it would have been there.

Christopher Columbus also thought the Earth was round. He was sure he could sail west from Europe and go all the way around the world to Asia. In 1492, Columbus set out on a trip around the globe. Unfortunately, he didn't make it. Instead, he bumped into America and ended his journey. Later, however, other explorers did sail all the way around the Earth and proved Columbus was right.

Columbus' Route 1492

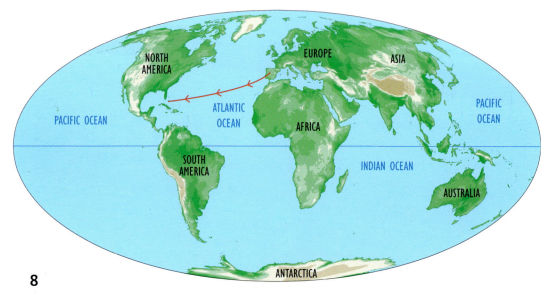

In recent years, travel to outer space finally proved without a doubt that the Earth is round. Pictures taken in space show that the Earth is indeed a sphere.

Scientist of Significance

John Glenn Jr. was the first American to orbit the Earth. Before becoming an astronaut, Glenn was an engineer and pilot with the U.S. Marines. In 1959, he became one of the seven men chosen as the first NASA astronauts. In 1962, he orbited the Earth three times as part of the Project Mercury space exploration crew.

A few years later, Glenn left the space program and became a U.S. senator. In 1998, he returned to space on the Discovery space shuttle. This time, at age 77, he became the oldest man to orbit the Earth.

Photo of the original seven Project Mercury astronauts
Front row (left to right): Walter M. Schirra Jr., Donald "Deke" K. Slayton, John Glenn Jr., and M. Scott Carpenter
Back row (left to right): Alan Shepard Jr., Virgil I. "Gus" Grissom, and L. Gordon Cooper Jr.

Tilted or Straight?

Find a globe in your home, classroom, or library. What do you notice about its position? Is it tilted or standing up straight? Have you ever wondered why globes are always tilted on their stands? Is it just a mistake by globe makers?

No, actually it's because the Earth is tilted on an **axis**. An axis is an imaginary line that runs through the middle of the Earth from the North Pole to the South Pole.

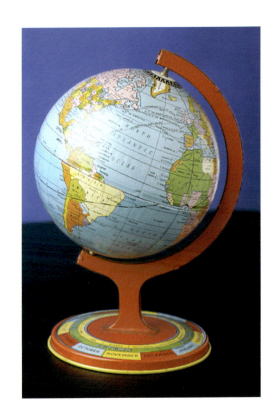

Follow That Star!

If you stretch the Earth's axis line up from the North Pole, it will point to the North Star. This star, also called the Pole Star or Polaris, is the only star in the sky that seems not to move.

People who live in the southern half of the globe look to the Southern Cross for direction. This constellation is a group of five stars, four of which form a cross.

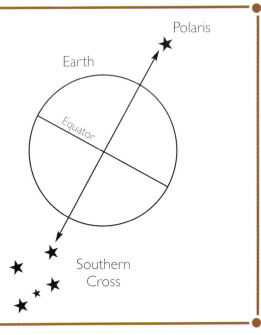

Polaris

Earth

Equator

Southern Cross

3

A Trip Around the Sun

"Sit still!"

How many times has someone said that to you? But it's hard to sit still. In fact, it's impossible! Earth is zooming around the Sun at about 18 miles per second. You can't help but go along. So the next time someone tells you to "sit still," you can tell them it's just not possible. (Of course, your parents and teachers might not appreciate that!)

But if you're moving so far so fast, why aren't you hanging on for dear life? In fact, it doesn't seem as if you're moving at all. That's because your body can only sense a change in speed or direction when there is a large change in motion. The Earth's movement isn't large enough for you to feel it.

A Long Journey to Understanding

Because the Earth's movement isn't felt, people went thousands of years believing that the Earth stood still. They thought that the Sun, Moon, stars, and other planets moved around the Earth. It took centuries and the work of many scientists before it became widely accepted that the Earth was moving.

Two scientists who were important to the discovery of the Earth's movement were Nicolaus Copernicus and Galileo Galilei. In the early 1500s, Copernicus wrote a book about his idea that the Earth and other planets revolved around the Sun. Most people of the time, however, didn't believe his claims. Unfortunately, Copernicus died before he could prove his theories.

About a century later, Italian **astronomer** Galileo backed up Copernicus's ideas. Galileo was the first scientist to use a telescope to look at the sky. His findings supported the idea that the Earth **revolved** around the Sun. However, at the time, the Catholic Church was very powerful, and Galileo's ideas went against church teachings. When Galileo published his work for everyone to read, the church forbid Galileo to leave his house for the rest of his life.

In time, the ideas of these two men were proven true. Their theories became the basis of understanding the Earth's movement today.

Time for a Trip

So how many trips have you made around the Sun? If you know how old you are, then you know the answer. You make one trip around the Sun each year. So your age is the total number of trips you've made around the Sun.

How long were these trips? Each one was 584 million miles. If you multiply this by your age, you'll find that you've covered a lot of miles in your lifetime!

Earth's trips around the Sun are called **revolutions**. One revolution takes about 365 days, or one year. Actually, each trip is about 365 ¼ days. But ¼ of a day wouldn't work on a calendar very well. So one extra day is added to the calendar every four years. This fourth year is called a *leap year*. Every leap year, February has 29 days instead of the usual 28.

Nothing Is Perfect

The Earth's path, or **orbit**, around the Sun is not a perfect circle. It is an ellipse, which is similar to an oval. At times, the Earth is farther from the Sun than at other times.

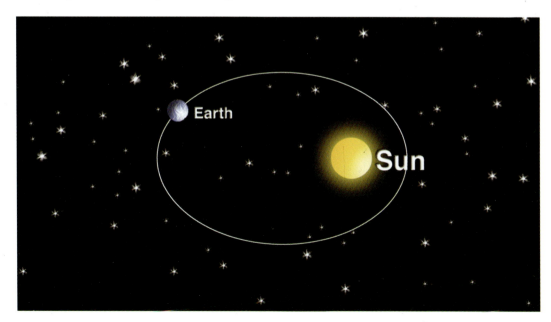

Spinning in Circles

The Earth doesn't just move around the Sun. It also spins on its axis like a gigantic top. This movement is called the Earth's **rotation**. The Earth makes one complete turn every 24 hours, or once a day.

Day and Night

The Earth's rotation causes day and night. As you rotate with the Earth toward the Sun, the Sun rises and day begins. As you rotate away from the Sun, the Sun sets and nighttime arrives. So when it's lunchtime where you live, it's the middle of the night on the opposite side of the world. And when you're snoring away at night, students in Asia are eating lunch.

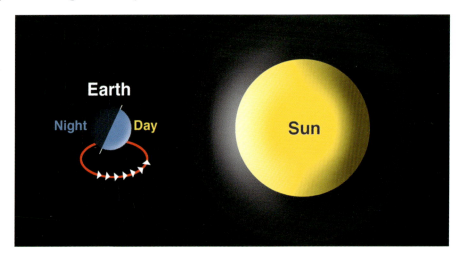

Earth

Night   Day

Sun

Me and My Shadow

As the Earth rotates, the Sun appears in different positions in the sky. These varying positions create changes in shadows. A shadow is a dark shape on a surface caused by an object blocking sunlight. The dark area takes the shape of the object. When you stand between the Sun and the ground, you block the Sun's rays, creating a dark image of your body behind or in front of you.

In the morning, the Sun is low in the sky. When the Sun is low, shadows are long. As the Sun moves higher in the sky,

shadows get shorter. When the sun is directly overhead, shadows are at their shortest. As the sun sets, shadows grow longer again.

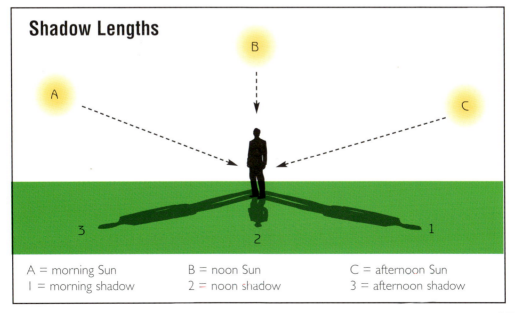

Shadow Lengths

A = morning Sun
1 = morning shadow

B = noon Sun
2 = noon shadow

C = afternoon Sun
3 = afternoon shadow

Inquire and Investigate
Shadows

Question: How does an object's shadow change during the day?

Answer the question: I think an object's shadow _____.

Form a hypothesis: An object's shadow _____.

Test the hypothesis:

Materials

- sidewalk chalk
- tall object such as a flagpole, lamppost, or street sign

Procedure

On a sunny day, locate a tall object that you can observe throughout the day. At 8 a.m., use the chalk to trace the object's shadow. Write the time by the shadow outline. Return to the object every two hours and trace the shadow. Label each outline. Continue doing this until the Sun sets. Then compare the lengths of your shadow outlines.

Observations: The shadow grew shorter in the morning and then increased in length in the afternoon.

Conclusions: An object's shadow decreases in length as the Sun rises and increases in length as it sets.

The Reasons for the Seasons

Have you ever taken a vacation to a warm tropical island in the middle of winter to escape the cold weather? Did you ever stop to wonder why it was warm and sunny on the island and not where you live?

The Earth's revolution around the Sun on its tilted axis is the reason for different seasons. As the Earth moves around the Sun, there is a certain time of year when part of it leans toward the Sun while part leans away. The part leaning toward the Sun has summer. The part leaning away has winter. Six months later, it is just the opposite. For the times in between, one part of the Earth is having spring while the other is having fall.

The United States is in the northern half, or Northern Hemisphere, of the Earth. In summer, the Northern Hemisphere tilts toward the Sun. The Sun shines more directly on the hemisphere, causing more hours of daylight and warmer temperatures. In winter, the hemisphere is tilted away from the Sun. The Sun's rays shine less directly on the hemisphere, so there are fewer hours of daylight and the temperature is colder.

When you're on winter break in the Northern Hemisphere, students are on summer vacation in the Southern Hemisphere.

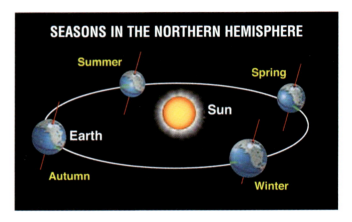

SEASONS IN THE NORTHERN HEMISPHERE

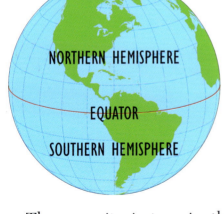

The opposite is true in the Southern Hemisphere. When the Northern Hemisphere is having summer, the Southern Hemisphere is having winter.

Hot or Cold

The Earth's tilted position also affects the type and length of seasons that occur in certain areas of the globe. The line that circles the middle of the Earth is called the *equator*. The area near the equator doesn't experience as much change in the directness of sunlight or in the number of daylight hours. So people who live near the equator don't have seasonal changes like those in other areas of the world.

The farther you move north or south from the equator, the greater the seasonal changes. However, as you reach the poles, the temperatures get colder for longer periods of time. This is because the Sun appears low in the sky at the poles, so heat from the Sun is less concentrated.

NORTH POLE

ECUADOR

Located about ten miles from the equator, Quito, Ecuador, has a mild, springlike climate all year long.

Inquire and Investigate
Seasonal Sunlight

Question: How does sunlight hit the Earth in its tilted position?

Answer the question: I think sunlight _____.

Form a hypothesis: Sunlight hits the Earth _____.

Test the hypothesis:

Materials

- blank wall
- flashlight

Procedure

Stand a few feet from the wall. Aim the flashlight straight at the wall. Observe the light brightness and pattern on the wall. Keeping the flashlight the same distance from the wall, tilt it so the light hits the wall about a foot above the first spot. Observe the light on the wall. Tilt the flashlight again so it hits about a foot higher. What do you notice about the light now? Try several more positions both above and below the straight-on point. What happens to the light?

Observations: The spot of light straight on was smaller and brighter. As it moves higher or lower, the spot spreads out and gets dimmer.

Conclusions: Sunlight hits the Earth most directly (straight on) at the equator. The farther from the center you move (toward the poles), the less direct and concentrated the sunlight is.

You're a Space Traveler!

You're spinning in circles while circling the Sun at the same time. You're zooming through space on the big, beautiful, blue planet called Earth. Maybe you've dreamed of becoming a space traveler. Guess what? You already are one! And you never even had to leave home.

Internet Connections for the Earth

http://www.allaboutnature.com/subjects/astronomy/planets/earth/
This simple site introduces the Earth, including its characteristics, tilt, and seasons.

http://www.windows.ucar.edu/tour/link=/earth/earth.html
Tour the Earth's interior, surface, atmosphere, and more with these basic facts.

http://www.dustbunny.com/afk/planets/earth/
Easy-to-read facts and photos of the Earth explain astronomy for kids.

http://kidsastronomy.com/earth.htm
Find fast facts about the Earth at this kids' astronomy site.

http://kids.msfc.nasa.gov/Earth/
Explore the Earth at this NASA student site.

http://www.scienceu.com/observatory/articles/seasons/seasons.html
The "reasons for the seasons" will be clear after you see the animated diagrams and read the information at this site.

Related Reading for the Earth

The Earth by Cynthia Pratt Nicolson. A unique combination of facts, folklore, simple experiments, and hands-on activities explains scientific discoveries about our planet. Kids Can Press, 1997. [RL 4.5 IL 2–5] (3217801 PB 3217802 CC)

Earth, Sun, and Moon by Robin Birch. This book is an introduction to the Earth, Sun, and Moon, describing what they are and how they move through the Solar System. Chelsea House, 2003. [RL 3 IL K–2] (6870206 HB)

Experiments with Gravity by Salvatore Tocci. Investigate gravity with these fun experiments. Children's Press, 2002. [RL 3.5 IL 3–5] (6872201 PB 6872206 HB)

Sunshine Makes the Seasons by Franklyn M. Branley. Describes how sunshine and the tilt of the Earth's axis are responsible for the changing seasons. HarperCollins, 1985. [RL 2 IL K–4] (8746401 PB 8746402 CC)

What Is Gravity? by Lisa Trumbauer. This book describes what gravity is and gives some examples of how it affects objects. Children's Press, 2004. [RL 1.2 IL K–3] (3514301 PB 3514302 CC)

What Makes Day and Night by Franklyn M. Branley. An excellent science book for beginning readers that explains the revolution of the Earth. HarperCollins, 1986. [RL 2 IL K–3] (9838301 PB 9838302 CC)

You're Aboard Spaceship Earth by Patricia Lauber. Earth is like a spaceship—it has everything on board that we need to survive: water, food, and air with oxygen. HarperCollins, 1996. [RL 2.5 IL 1–4] (4955201 PB 4955202 CC)

•RL = Reading Level
•IL = Interest Level
Perfection Learning's catalog numbers are included for your ordering convenience. PB indicates paperback. CC indicates Cover Craft. HB indicates hardback.

Glossary

astronomer (uh STRAHN uh mer) person who studies objects in the sky (Sun, Moon, planets, stars, etc.)

atmosphere (AT muhs sfear) mixture of gases that surround a planet or other body in space

axis (AK sis) imaginary line around which an object turns

galaxy (GAL uhk see) group of stars, planets, gases, and dust

gravity (GRAV uh tee) force of attraction between two bodies in space

orbit (OR bit) to move around a body in space in a path controlled by the force of gravity (verb); path that a body in space follows around a larger body (noun) (see separate entry for *gravity*)

propellant (proh PEL ent) material that's burned to give upward thrust to a rocket

revolution (rev uh LOO shuhn) circular movement of one object around another

revolve (ree VAWLV) to move in a circular motion around an object

rotation (roh TAY shuhn) spinning motion around an axis or fixed point (see separate entry for *axis*)

shadow (SHAD oh) dark shape on a surface that falls behind an object blocking a source of light

sphere (sfear) round object shaped like a ball

23

Index